Talita Sarah Mazzoni
Maria Dalva Cesario
Angelina Batista

Teaching embryology through the construction of educational materials

Talita Sarah Mazzoni
Maria Dalva Cesario
Angelina Batista

Teaching embryology through the construction of educational materials

Models and games for teaching embryology in schools

ScienciaScripts

Imprint

Cover image: www.ingimage.com

This book is a translation from the original published under ISBN 978-3-330-76527-6.

Publisher:
Sciencia Scripts
is a trademark of
Dodo Books Indian Ocean Ltd. and OmniScriptum S.R.L publishing group

120 High Road, East Finchley, London, N2 9ED, United Kingdom
Str. Armeneasca 28/1, office 1, Chisinau MD-2012, Republic of Moldova, Europe
Managing Directors: Ieva Konstantinova, Victoria Ursu
info@omniscriptum.com

Printed at: see last page
ISBN: 978-620-8-51975-9

CONTENTS

CHAPTER 1	4
CHAPTER 2	9
CHAPTER 3	23
CHAPTER 4	26
CHAPTER 5	38
CHAPTER 6	45

PREFACE

Biology can be taught in schools in a way that is full of experiments, practical and dynamic lessons, but this doesn't often happen due to the extensive syllabus and the reduced workload. The most memorable biology lessons for any student are those where there is a break from routine, whether it's a science fair, practical lessons or lectures with different resources from those used in a lecture. For this reason, paradidactic materials appear as a necessary resource that facilitates learning, making it meaningful and not based purely on memorising the content. Students often find it difficult to imagine an object in three dimensions and fail to understand the dynamic processes that occur in organisms.

In order to reduce these limitations and involve the student in the learning process, it is important to use resources capable of attracting the student's attention and strengthening the retention of the concepts taught. With this in mind, the work presented here aims to provide ideas for practical biology lessons through the construction of paradidactic instruments made from low-cost materials.

The aim of this project was to design, develop and apply a paradidactic material consisting of a game and models for secondary school students, covering the uterine cycle, the ovarian cycle, fertilisation, the cleavage process and the implantation of the blastocyst in the uterus, clarifying the morphological changes that occur during the female sexual cycle, as well as the hormonal changes present during these processes in human beings.

The models made in this work using plastic, papier-mâché and biscuit dough, together with the game, have been used since 2005 to date during secondary school lessons and have provided positive responses in the teaching-learning process. Making the models and the game available to schools through temporary loans has made this constant application possible over the years.

For demonstration purposes, the first results obtained in 2005 will be described here, based on the application of the paradidactic material to students in the first year of secondary school at the "Professor Américo Virgínio dos Santos" State School, in the city of Botucatu-SP.

However, over the years it has been possible to observe that learning and student participation are always much more satisfactory when these resources are used, which attract the students' attention and reinforce the retention of the concepts learnt. The use of games always provides stimulation and a favourable environment for the development of interpersonal relationships, communication and expression, leadership and teamwork, showing students a new playful, enjoyable and participatory way of relating to school content.

Thus, in a simple way and with a little creativity, teachers can create valuable elements in the process of appropriating knowledge, making their lessons more dynamic and attractive, involving their students more meaningfully in the teaching-learning process. In addition, the use of models and games made by the teacher themselves, and perhaps by the students, makes them easily accessible, unlike other more sophisticated teaching resources that are difficult to access in schools that don't have them. We therefore believe that these paradidactic resources can stimulate and facilitate the teaching-learning process, making it more dynamic and attractive for students.

CHAPTER 1

Introduction

Biology, as part of the development of humanity, has the following teaching objectives: to provide an understanding of man as a living being and his role in nature; to give an insight into the variety of life and the similarities between living beings; to teach the art of planning scientific investigations, formulating questions and organising experiments; to develop a spirit of criticism of the evidence and to provide an up-to-date view of the content (Krasilchik, 1983). However, it can be seen that in recent years there has been a decline in the quality of biology teaching in schools.

Students are not encouraged to discuss the causes of phenomena or establish causal relationships, trying to understand the mechanisms of the processes they are studying (Krasilchik, 1987). This is due to a very extensive syllabus associated with a reduced workload and often because there is a deficiency in teacher training. This lack of professional training leads to a dependence on textbooks , poor quality lessons and insecurity in relation to the class, because they don't master the subject (Krasilchik, 1987).

Biology, as a science, makes it possible for teaching in schools to be full of experiments, practical and dynamic lessons, but this doesn't happen very often. Undoubtedly, the most memorable biology lessons for any student are those in which there is a break from routine, whether it's a science fair, a practical lesson or a lecture with different resources from those used in a theoretical lesson. Although biology teachers themselves recognise the importance of practical lessons, they are far too few because there is not enough time to prepare the material, there is a lack of security to control the class, there is a lack of knowledge to organise experiments and there is a lack of suitable equipment and facilities (Krasilchik, 2004).

In general, teachers tend to adopt the traditional teaching model, in which learning takes place through the mere transmission-reception of so-called "scientific truths" (Campos & Nigro, 1999). The more traditional teaching models are based on the textbook as the element that shapes the programmes and mediates the relationship between the teacher and the students. This type of material corresponds to teaching centred on basically transmissive models and based on conceptual content (Coll et al., 2003). In addition, textbooks have limitations and shortcomings and, however good their quality may become, it is essential to use paradidactic materials, which need to be more present and systematic in school education (Delizoicov et al., 2002). More than necessary, these resources should be incorporated into everyday school life, as they are essential to the teaching-learning process, awakening students' skills, creativity and rational thinking, developing a dynamic and dimensional view of the world and thus expanding their cognitive possibilities.

According to the Parâmetros Curriculares Nacionais do Ensino Médio (PCNEMs - National High School Curriculum Parameters), students who arrive at high school bring with them a wealth of biological knowledge acquired at different educational institutions. However, this knowledge is almost always fragmented and incomplete. The content to be taught, the use of methodological principles and the choice of appropriate procedures helps to broaden the general view of the world and guarantees learning. Of course, teaching cannot be limited to lectures. There should be debate and reflection between teacher and student, as well as the use of complementary resources for understanding reality, such as games, projections and exhibitions. These resources must be used in such a way that the student interacts with the object of study, enabling them to analyse and criticise it.

In a constructivist conception of learning, understanding teaching as empowering all of a person's capacities, the conception of teaching must take into account the diversity of students, in which the teacher's role consists of presenting challenges and providing the appropriate help for each student's needs. In this context, educational materials appear as a necessary resource and facilitator of learning (Coll et al., 2003), making it meaningful and not based purely on memorising content.

It should be emphasised that educational material is not only used to illustrate and facilitate the understanding of basic concepts. Much more than that, it is an important tool capable of stimulating, instigating and developing students' logical and abstract thinking.

Students often find it difficult to imagine an object in three dimensions and are unable to understand the dynamic processes that occur in organisms (Krasilchik, 1983). In order to reduce these limitations and involve the student in the learning process, it is important to use resources capable of attracting the student's attention and strengthening the retention of the concepts taught. According to Harlan & Rivkin (2002), the pleasure of being involved in the preparation or development of practical activities provides an intense emotional state in relation to lasting memories, reinforcing the retention of concepts.

Another source of great difficulty in learning biology is the lack of intra-disciplinary integration. The content is presented divided into isolated compartments, without providing students with opportunities to synthesise and give coherence to the whole, which would be done by showing them the links between facts, phenomena, concepts and processes learned (Krasilchik, 1983). As a result, some important areas of biology have been devalued in secondary education.

Embryology, the science that studies the origin and development processes of a new living being, from fertilisation to the formation of a highly complex organism (Garcia et al., 2011), is an example of this. The transformation of a single cell, resulting from the union between sperm and egg, into a complex organism should be, at the very least, a source of great curiosity, beauty and cultural pleasure, so learning about it is simply fundamental (Osse, 2000). However, following the pattern of biology teaching, the teaching of embryology is divided into several areas and is often relegated to the background, leaving little space for its teaching or even becoming totally absent from the school curriculum due to lack of time. This situation of compartmentalising subjects makes it difficult for students to understand that biology is a comprehensive and integrative science that should be seen as a whole, where there are correlations between the most varied subjects.

It has been observed in secondary schools that the content of embryology only covers the processes of cleavage, gastrulation and embryonic appendages, and is dealt with within the subject of zoology, leaving the other subjects involved to other areas of biology. Subjects such as gametogenesis and fertilisation are covered in genetics, while hormonal control in the sexual cycle is covered in physiology, within the endocrine or reproductive system.

Thus, considering that: a) school content is a reflection and selection of those aspects of culture whose learning is considered to contribute to students' social and individual development (Coll et al., 2003); b) that the subjects covered by embryology are extremely important and should never be undervalued and; c) that when studying the patterns of animal reproduction, models should be built that allow a spatial view of the changes that the egg cell undergoes in its process of segmentation and cell differentiation (Brasil, 1998), the aim of this study was to design, develop and apply a paradidactic material consisting of a game and models for teaching embryology in secondary schools. From the application of the material developed in the classroom, which illustrates the uterine cycle, the ovarian cycle, fertilisation, the cleavage process and the implantation of the blastocyst in the uterus, it is hoped to clarify the morphological changes that occur during the female sexual cycle and correlate them with the hormonal changes present during these processes in human beings. It is also hoped that the use of the models in the classroom, combined with lessons on the aforementioned content, will spark debate and reflection between teachers and students on social issues such as the formation and development of a new individual, sexuality, teenage pregnancy, contraceptive methods and more controversial and current issues such as assisted reproduction, the use of stem cells in cell therapy, abortion, cloning, when life begins, in short, useful and important knowledge that can contribute to students' personal and social development.

CHAPTER 2

Procedure

2.1. Mock-ups

- Female gamete - Ovocyte

The model of the female gamete was created based on the mature oocyte, which undergoes ovocation and is released by the ovary along with the other layers that surround it (zona pellucida and corona radiata) and the 1st polar body (Figure 1a and 1b). At the time of ovocation, the oocyte is in metaphase II of meiosis (Figure 1c), which was also represented in the model.

The oocyte and the 1st polar body were made from Styrofoam balls.

The oocyte was made from a Styrofoam ball 100 mm in diameter, cut in half to show the inside of the oocyte, where the chromosomes in metaphase II of meiosis were represented with small coloured plastic threads. The inside of the oocyte was moulded with biscuit dough, which can be found at Annex I. The centrioles were made from rubber and the spindle fibres were drawn with a pen in silver ink. The oocyte was coloured with blue paint.

The layers surrounding the oocyte were made using a hollow Styrofoam ball 150 mm in diameter. The corona radiata, the cell layer, was represented by Styrofoam balls 5 mm in diameter, which were cut in half and glued to the outer surface of the Styrofoam ball. Thus, each Styrofoam ball represents a cell of the corona radiata. After gluing the balls together, this surface, representing the corona radiata, was painted lilac. The zona pellucida, the glycoprotein layer, was represented by the inner edge of the 150 mm Styrofoam ball, which was painted light green. The inner space of the 150 mm Styrofoam ball was painted yellow, representing the perivitelline space containing the oocyte and the 1st polar body. To make the 1st polar body, a 25 mm diameter Styrofoam ball was used, which was painted dark green.

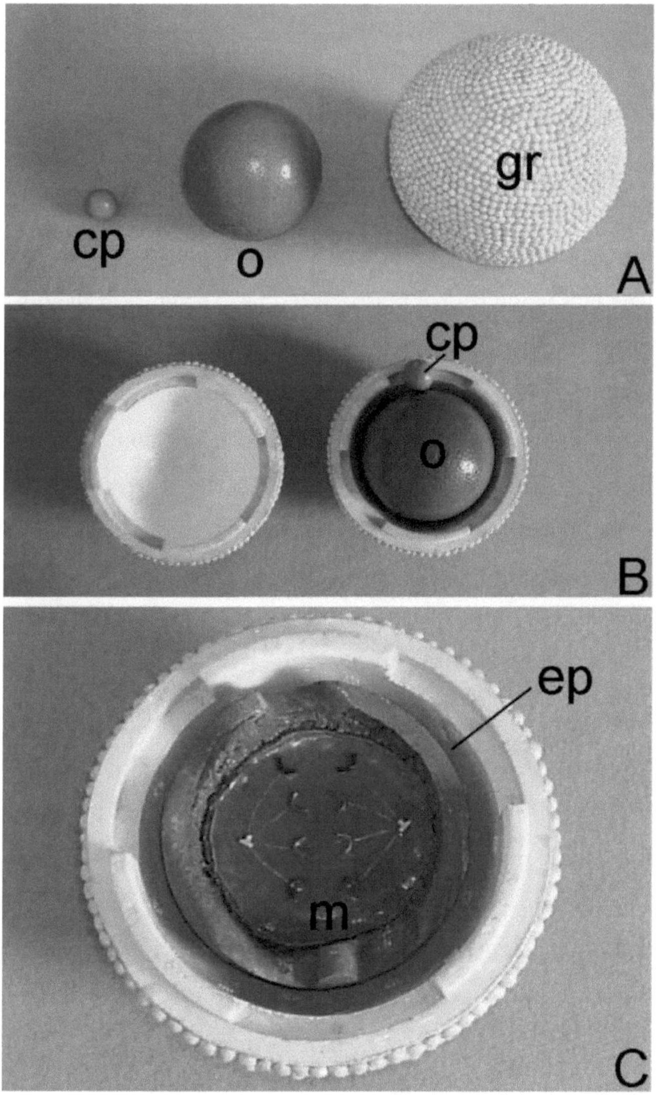

Figura 1: **Female gamete. A)** Mock-ups of the 1st polar body (cp), the oocyte (o) and the corona radiata (granulosa - gr). **B)** showing the 1st polar body (cp) and oocyte (o), in the perivitelline space, surrounded by the zona pellucida and corona radiata. **C) Mock-up** of the female gamete, showing the interior of the oocyte, which is in metaphase II of meiosis (m). Perivitelline space (ep).

- **Male gamete - Spermatozoon**

The sperm model (Figure 2a) was made following the morphological structure of the gamete.

A transparent plastic soap dish was used for the sperm head and the acrosome-covered nucleus was placed inside (Figure 2b). The nucleus was made from an oval piece of Styrofoam covered in biscuit dough. Once ready, the acrosome was moulded with biscuit dough around and on top of the nucleus (Figure 2c). Plastic was placed between the nucleus and the acrosome to prevent the acrosome from adhering to the nucleus, so that the acrosome would remain mobile in the finished model (Figure 2d).

The intermediate piece was made from a plastic test tube. The ends of the tube were cut off and a coloured telephone wire was wound inside, representing the mitochondria present in this region (Figure 2b). The test tube was glued to the soap dish and a plastic hose was glued to its end, representing the main and final parts that make up the sperm flagellum. A wire was placed inside this hose, representing the microtubules that support the flagellum. This wire also made this part flexible, allowing the tail of the sperm to be positioned as the teacher saw fit.

- **Phases of fertilisation**

Four phases of fertilisation were represented:

O Sperm inside the cytoplasm of the oocyte and release of the 2nd polar body (Figure 3a)

F Formation of male and female pro-nuclei, 1st and 2nd polar bodies (Figure 3b)

Fusion of the male and female pro-nuclei, 1st and 2nd polar bodies (Figure 3c)

F Formation and division of the zygote, showing the chromosomes during cell division and the polar bodies in the perivitelline space (Figure 3d)

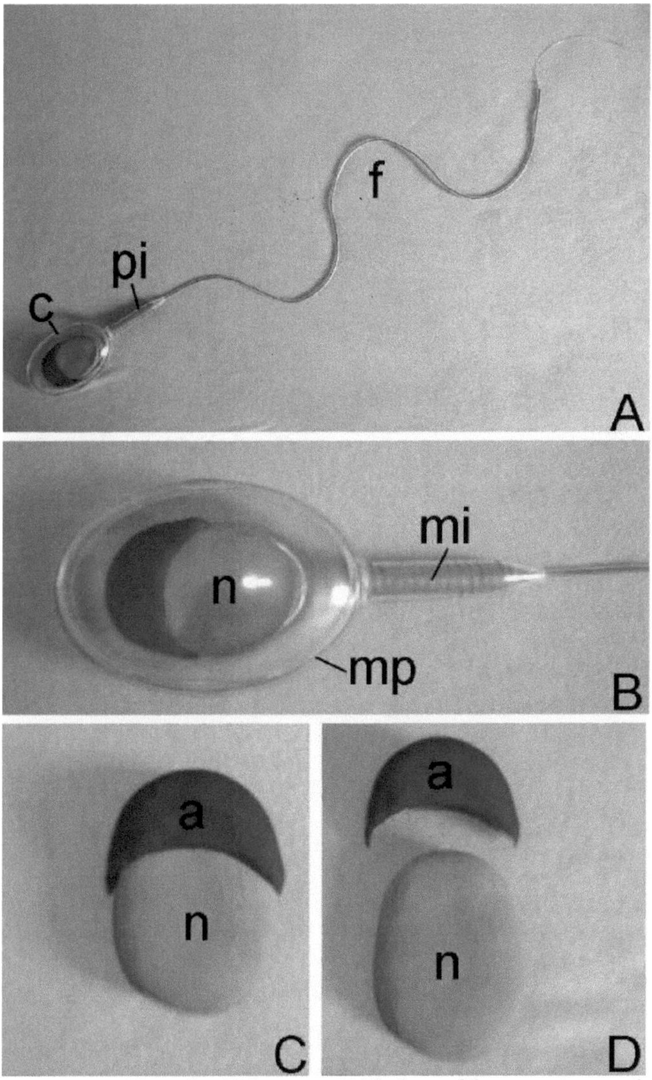

Figura 2: Male gamete. A) Mock-up of the spermatozoon, showing the head (c), midpiece (pi) and flagellum (f). **B)** Detail of the sperm mock-up showing the head region, with the nucleus (n), plasma membrane (mp) and the intermediate piece with mitochondria (mi). **C) Mock-up** of the sperm nucleus (n) covered by the acrosome (a). **D) Mock-up** of the sperm nucleus (n) separated from the acrosome (a).

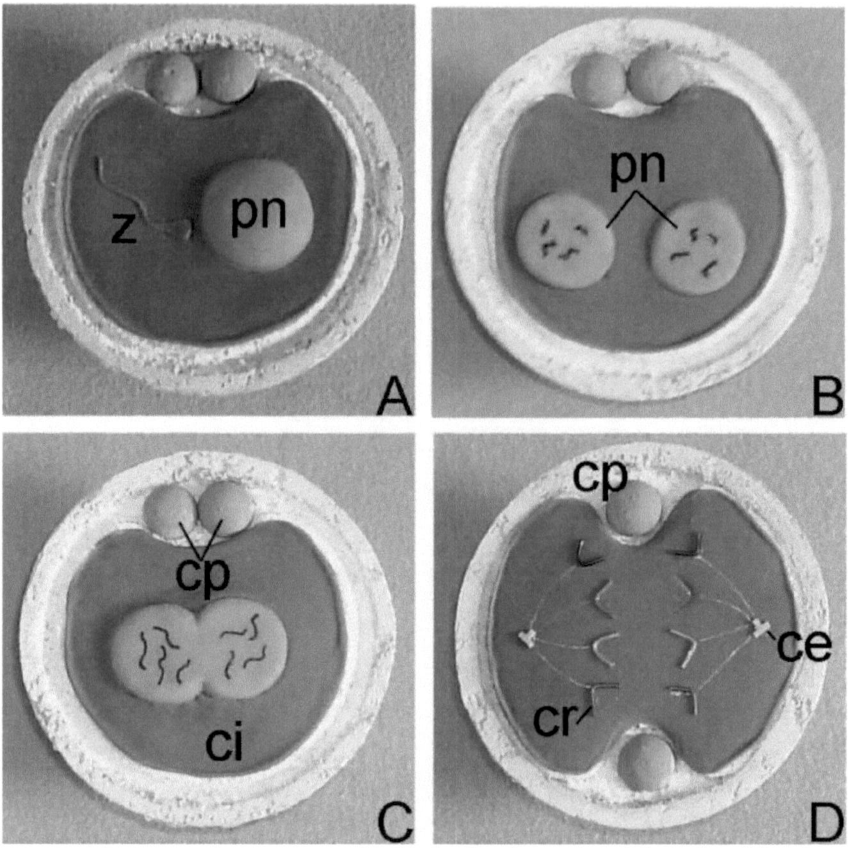

Figura 3: Phases of fertilisation. A) Model showing the meeting of the male and female gametes. **B) Mock-up** showing the formation of the male and female pro-nuclei. **C) Mock-up** showing the fusion of the male and female pro-nuclei. **D) Mock-up** showing the formation of the zygote and the first mitotic division. Pro-nucleus (pn), sperm (z), cytoplasm (ci), chromosome (cr), centriole (ce).

Each stage was made using a Styrofoam ball 100 mm in diameter cut in half. Shown are the oocyte and the polar body in the perivitelline space, surrounded by the zona pellucida.

The polar bodies were made from Styrofoam balls 20 mm in diameter, cut in half. These were painted dark green and glued to the perivitelline space.

The oocyte, the sperm and the male and female pro-nuclei were moulded on Styrofoam with biscuit dough and painted blue, brown and pink, respectively.

To represent the perivitelline space, the Styrofoam ball was hollowed out using sandpaper. The space was then painted yellow.

The chromosomes of the pro-nuclei were drawn with a felt-tip pen on the biscuit dough and the chromosomes of the cell division were represented with the same material used to make the chromosomes of the female gamete.

The outer surface of the Styrofoam ball was painted light green, representing the zona pellucida.

- Female Reproductive System

This model shows only half of the reproductive system (Figure 4a). The uterus and fallopian tube are shown in a longitudinal section, while the ovary and infundibulum with the fallopian tube fimbriae are shown in their entirety.

Firstly, the reproductive system was drawn on a 30 mm thick Styrofoam plate. The drawing was then cut out to represent the outer and inner walls of the uterus and fallopian tube and fixed onto another 30 mm thick Styrofoam plate. The mould of the reproductive system was then covered and modelled with papier-mâché (described in Appendix II). Once dry, the material was painted various shades of red and pink.

To represent the fimbriae, red crepe paper was used, cut out in the shape of fringes and attached to the end of the fallopian tube (Figure 4b).

The broad ligament of the uterus was represented in the model by the Styrofoam area between the uterus and the ovary, which was painted yellow.

The Styrofoam plate on which the model was fixed was painted blue.

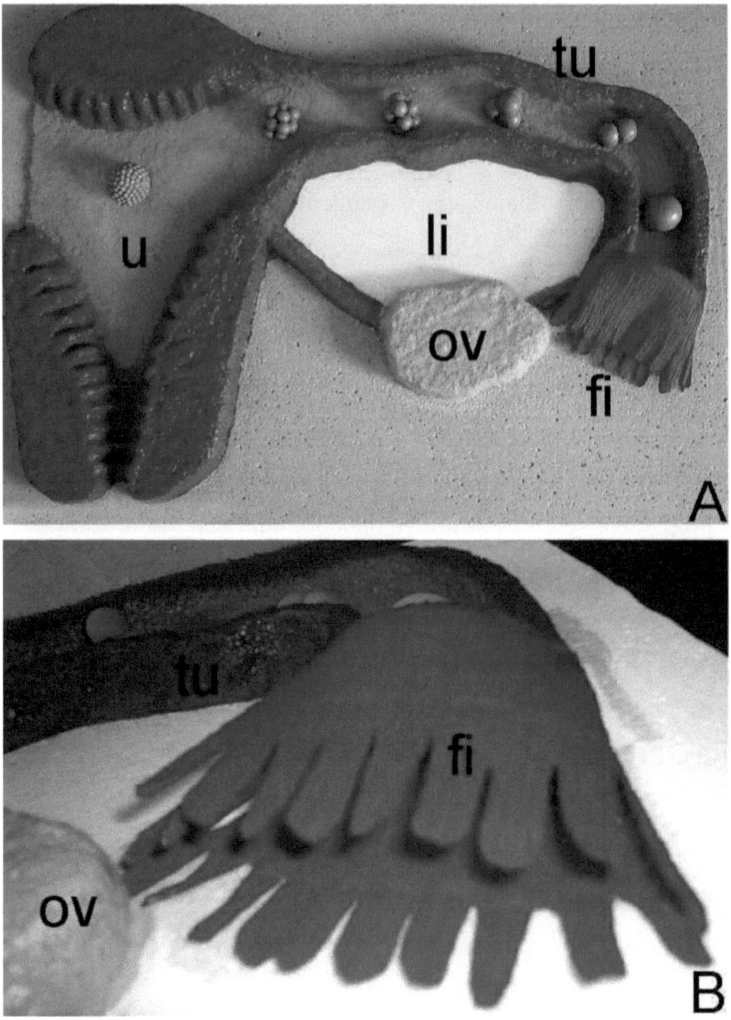

Figura 4: Female Reproductive System. A) Model of the female reproductive system, showing the cleavage process inside the fallopian tube. **B)** Detail of the model showing the region of the infundibulum with the fimbriae. Uterus (u), broad ligament (li), fallopian tube (tu), ovary (ov), fimbriae (fi).

- Cleavage

The cleavage process was represented from the zygote to the formation of the blastocyst (Figure 5a). For this purpose, Styrofoam balls of different sizes were used and coated with biscuit dough. The balls, representing the blastomeres, were joined together with the biscuit dough itself.

The zygote was made with a Styrofoam ball 35 mm in diameter. The two-blastomere stage, with two Styrofoam balls 25 mm in diameter. The four-blastomere stage, with four Styrofoam balls 20 mm in diameter. The eight-blastomere stage, with eight Styrofoam balls 15 mm in diameter. The morula was made with 16 Styrofoam balls 10 mm in diameter. To represent the blastocyst cells, Styrofoam balls 5 mm in diameter were cut in half and glued onto a ping pong ball 35 mm in diameter.

After the biscuit dough had dried, all the material was painted with blue paint.

At this stage, the zona pellucida was represented as follows: each of the cleavage stages, from the zygote to the morula stage, was placed inside a transparent plastic latex bladder (Figure 5b). However, as latex breaks easily, the material was prepared at the time of application, in the classroom.

- Blastocyst

The blastocyst is a structure made up of an outer layer of cells called the trophoblast and a group of cells arranged internally, called the embryoblast or inner cell mass. To represent it, we used a transparent, hollow plastic ball that could be opened into two halves. In order to show the inside of the blastocyst, a model was made depicting the blastocyst in sagittal section (Figure 5c). On the inside of each half of the plastic ball, the internal cell mass was represented, made with biscuit dough, covered with Styrofoam balls 5 mm in diameter, cut in half. On the outside of the transparent ball, these same balls cut in half were also used (Figure 5c). The outer surface was then painted the same shade of blue as the mock-ups of the cleavage phases, while the inner cell mass was painted a light blue colour.

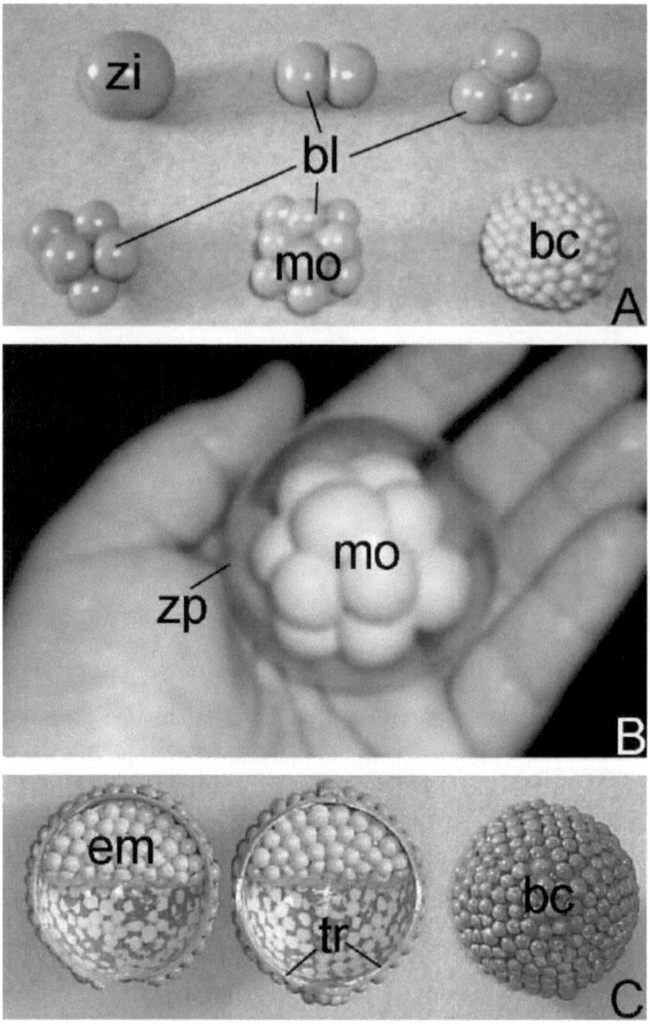

Figura 5: Phases of cleavage. A) Mock-ups of all the cleavage stages, from the zygote (zi) to the blastocyst (bc). **B)** Mock-up of the morula stage (mo) surrounded by the zona pellucida (zp). **C) Mock-up** of the blastocyst (bc), showing the embryoblast (em) inside and its outer surface covered by the trophoblast (tr). Blastomeres (bl).

All the material made was painted with gouache paint and once dry it was given a layer of white glue to protect the paint.

To make the models, we also used materials such as: a Styrofoam cutting machine, white glue, Styrofoam glue, hot glue, a stylus, paintbrushes, sandpaper, among others.

To protect the material and make it easier to transport, the model of the Female Reproductive System was stored in a wooden box and the rest of the models were separated into smaller boxes.

2.2 Game design

The game (Figure 6a) was designed on the basis of the content on the female sexual cycle found in secondary school textbooks and the existing literature on educational games, so that it was possible for all the students to take part in the game.

The game consists of:

6 6 of a kind

6 66 different questions

1 180 identical mini cards

2 216 different mini-cards, which make up 6 types of puzzle

The game was devised using the Bingo game as a reference. The game cards were printed on sturdy paper and then laminated with contact paper.

Each card (Figure 7) is made up of 30 different answers to the questions, of which 18 questions are present on all the cards, 6 questions are common to 3 different cards and 6 questions are specific to each card, making it difficult to tie the game. The answers on the cards are arranged in 6 rows and 6 columns, resulting in 36 quadrants, 6 of which are filled with figures that help complete the row or column. The questions, which must be drawn one by one by the teacher, have been printed on small cards (Figure 6b), which contain a single question and its respective answer.

- **Game rules**

The room is divided into 6 groups, each group receiving a card and a set of mini-cards. The teacher draws one question at a time and each group has to find the right answer on their card, overlaying each answer with a mini-card. Each question must be drawn only once. When a row or column is completely filled, the group has two options: choose a person from another group and bring them into their group or expel a participant, choosing another group to receive them. The group that fills in the whole card first is the winner.

The game has two play options. In the first option, the group simply fills in the card using the same mini-cards (Figure 6c). In the second option, the group receives 36 mini cards that make up a puzzle (Figure 8). The intention is that while filling in the card, the group realises that the mini-cards form a figure and tries to organise them, assembling the puzzle correctly. The group that manages to complete the puzzle should briefly explain what the figure represents in the female sexual cycle. To encourage students to take part, the teacher can offer some prizes for the winners, such as sweets, chocolates or whatever else they find interesting.

The questions drawn by the teacher should be placed on the table, keeping the order of the draw so that the winning group's card can be checked.

- **Game pre-assessment**

Before being applied in the school, the game was tested with the first year students of the Biomedicine course at Unesp, Botucatu-SP, after an Embryology class on the Female Sexual Cycle. The purpose of the test was to see if the game would work. After applying the game, some questions that were not clear to the students or that caused doubts were reformulated to be applied later in secondary school.

The game was placed in a plastic case to protect the material and make it easier to transport. The cards and mini-cards were placed in plastic bags, and the mini-cards were separated into two boxes, one containing the same mini-cards and the other containing the mini-cards that form the puzzle. The cards with the game questions were placed in a plastic box to allow them to be shuffled and drawn by the teacher.

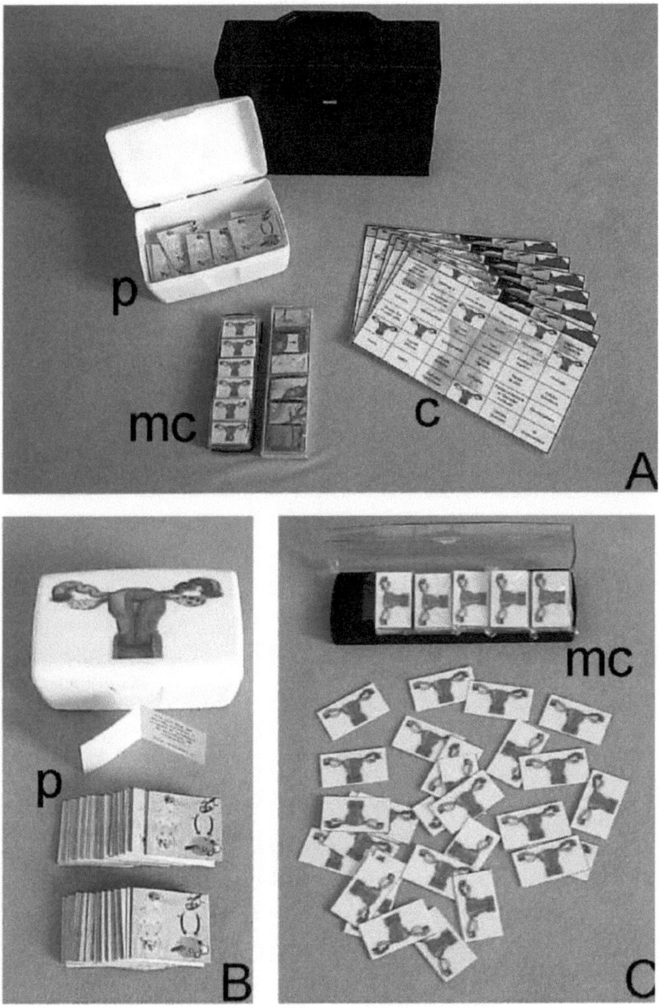

Figure 6: Game about the Female Sexual Cycle. A) Complete game showing the cards (c), micro-cards (mc) and question cards (p). **B)** Cards with the questions (p) to be drawn. **C)** Identical mini-cards (mc).

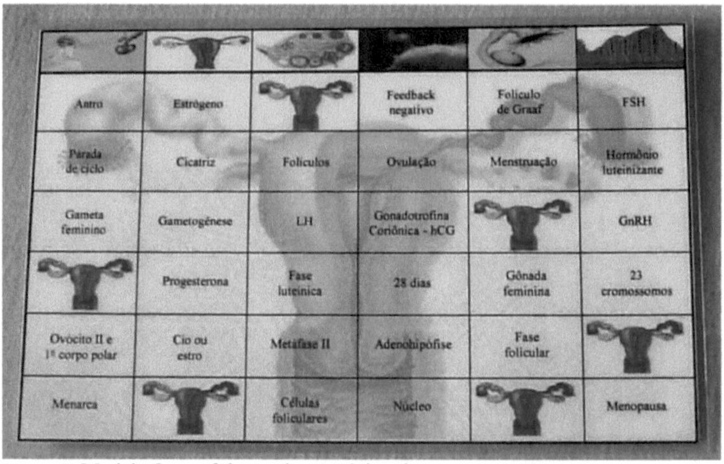

Figure 7: Model of one of the cards containing the answers to the game.

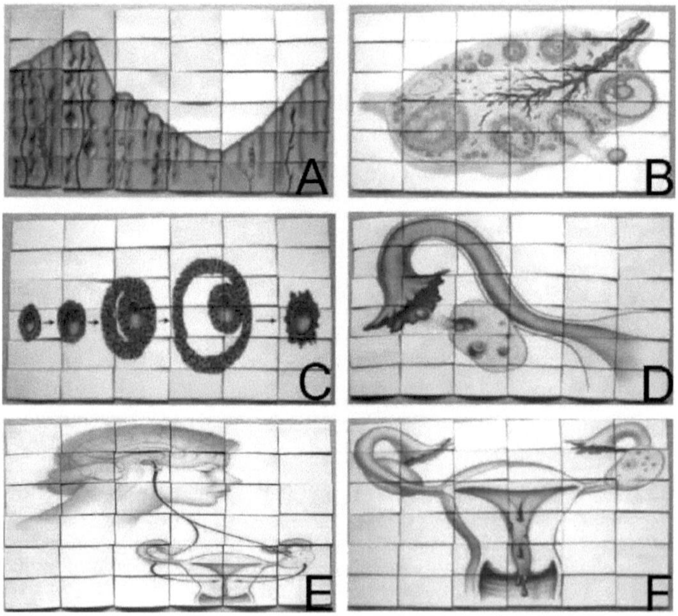

Figure 8: Different mini-charts. A-F) The 6 types of puzzle formed by the mini-charts with different illustrations of the female sexual cycle.

CHAPTER 3

Mitería application

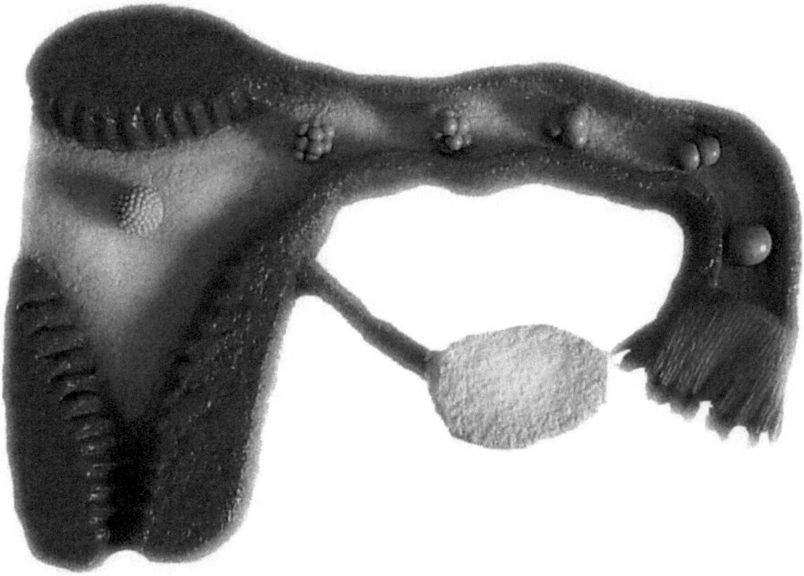

The educational material presented in this work has been used since it was produced in 2005. The kit with all the materials is available at the Department of Morphology at Universidade Estadual Paulista (Unesp - Campus de Botucatu, SP) and at the Department of Cell and Developmental Biology at Universidade Federal de Alfenas (Unifal - Campus de Alfenas, MG).

In order to demonstrate the results obtained, we will describe the application of the material in a single school, considering that the results in the teaching-learning process, after the application of the models and the game, are always very similar.

The material was applied for the first time to students in the first year of secondary school at the "Professor Américo Virginio dos Santos" State School in the city of Botucatu, in the interior of São Paulo.

The material was applied in a single day, using six consecutive hours in the form of a mini-course.

Initially, a lecture was given on the topics: Female Sexual Cycle, Fertilisation and Cleavage. To facilitate understanding, the lecture was conducted using a slide show, which was borrowed from the Morphology Department of the IB in Botucatu, UNESP.

After the lecture, the models of the Female Reproductive System, cleavage stages, male gamete, female gamete and fertilisation stages were presented and explained to the students. The students were asked to recognise the structures of each of the models, relating them to the subject covered during the lecture. To do this, the students were given the models so that they could examine them properly. The only model that remained on the teacher's desk was that of the Female Reproductive System, due to its size and fragility. To observe it, the students, divided into small groups, went to the teacher's desk and looked at the models, trying to place the concepts corresponding to the Cleavage stages in the correct position and order inside the tube and uterus.

To check learning, the game on the Female Sexual Cycle was played and then an assessment was carried out.

Thirty-nine students took part in the game, which was divided into six groups according to the affinity of the participants. There were two groups of six people, two groups of seven people, one group of four people and one group of nine people. Each group received a card and a set of mini-cards. The mini cards used were those that make up the puzzle. With each question drawn, the students answered the questions aloud and only then looked for the correct answer on the card. When a row or column was completely filled by a group, all its members received candies and could choose whether to expel or invite a member of another group to join them. Only the winning group remained with the same members, as they didn't want to expel or invite anyone throughout the game. All the groups realised that the mini cards formed a puzzle, but only two groups managed to form it correctly, including the winning group. They received a box of chocolates as their prize.

Afterwards, an assessment was carried out containing 10 multiple-choice questions on the subjects covered and an open question for the students to express their opinions on the lesson and the materials used (Appendix III).

CHAPTER 4

Results and Discussion

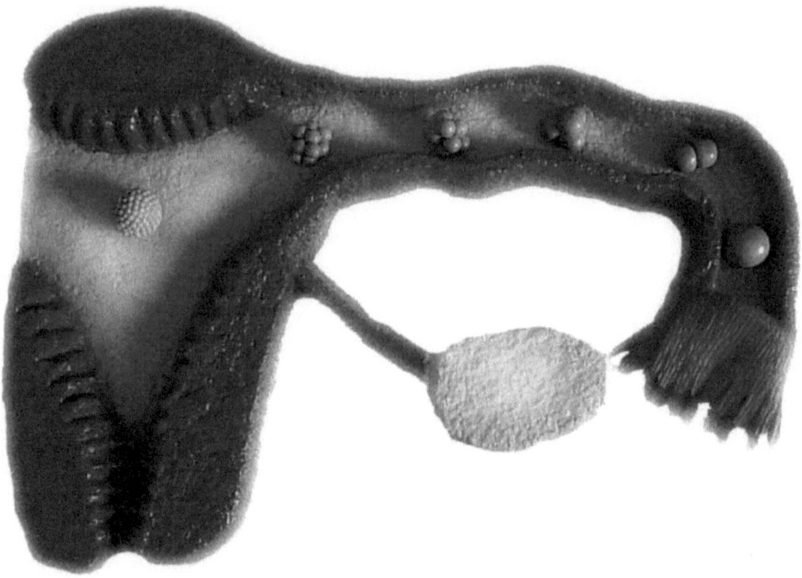

4.1. Lectures

During the lecture, the topics of the Female Sexual Cycle, Fertilisation and Cleavage could be covered in great detail because the time available was sufficient and the illustrations used on the slides helped the process. In addition, the subjects could be dealt with very calmly and were returned to several times so that the students could assimilate and understand all the new information they were receiving. The students remained silent most of the time, showing great interest, curiosity and surprise. When someone started to talk, the students themselves drew the attention of their classmates by asking them to be quiet.

During the slide show, many students were amazed by the pictures shown, especially those that referred to the inside of the human body, the female reproductive system and the male and female gametes. With each new slide, it was possible to hear expressions from some students saying: "look, that's beautiful ", "wow, that's cool", "I've never seen that", "it's just like the drawings".

According to Veiga (2003), the teacher should conduct an expository lesson on a given subject, offering the correct explanations on the topic covered, but should allow students to disagree, argue, question and expose their ideas, thus facilitating learning. This could be observed several times during the lesson. There were many questions during the explanation, especially when the Menstrual Cycle and Cleavage were discussed. The students were curious to know in which period of the Menstrual Cycle fertilisation occurs, why menstruation is interrupted during pregnancy, how twin siblings form, what stem cells are, how contraceptive methods work in the sexual cycle and when they should be used, how pregnancy tests bought at the pharmacy work, etc. It can be seen that, although these points were not addressed directly, the students were able to correctly correlate these current issues with the subject in question. Many misconceptions were corrected during the questions asked by the students, and many doubts were resolved.

The biggest difficulty faced during the presentation of the slides was the school's infrastructure. The first year classroom, in which the mini-course would initially be taught, had a very thin white curtain, which was unable to darken the room. Nevertheless, the lesson began. However, after 15 minutes, the sun became very strong and made it impossible to continue the lesson, as the slides were being projected on the blackboard, since there was no other suitable place. In an attempt to continue showing the slides, a white curtain was taped over the blackboard, but the crumpled cloth distorted the image and kept coming loose from the blackboard. So we had to change rooms with the fourth grade, as it was the only room suitable for slide projection. In the new room, the curtains were dark, but to make it easier to see the slides, the lights had to be switched off. This was another problem, as the room didn't have a light switch. The light was switched off at the main switch, which also switched off the socket that the slide projector was plugged into. This problem was solved by leaving the room unlit and connecting the slide projector to the socket in the classroom next door. This required the use of three extension cables. As this room also didn't have a projection screen and none of the walls were white, the slides were projected onto the blackboard. Although the blackboard was full of stains (figure 9a), it was possible to visualise the slides and continue the lesson. Another problem was positioning the projector so that the slides were projected clearly, at the ideal height and distance. To do this, the slide projector was placed on a chair, which was placed on top of two desks.

Although the content covered was quite extensive and complex for secondary school, only a few students found the lecture tiring. When the slide show ended, many students asked for more figures to be shown and said they would like to have more lessons in this way.

The lecture ended a few minutes before the school break. However, instead of going to the playground, most of the students remained in the classroom asking questions, bringing up facts from their daily lives, expressing their ideas and paying attention to the answers given. This small debate lasted throughout the break and some students even almost missed their lunch break.

4.2. Application of the models

As for the models, from the moment they were mentioned at the start of the lesson, the students were extremely eager to see them. So even before the break was over, most of the students had already returned to the classroom. At this point, it was difficult to contain the uproar in the room and get them to calm down and remain seated so that the lesson could resume.

During the observation of the models (figure 9b-f), many doubts could also be clarified. The models, used by the teacher as explanatory and representative models to explain the functioning, phenomena and certain biological processes that occur in an organism, facilitate learning and broaden the understanding of more abstract subjects, agreeing with the PCNEMs' statements.

The second part of the lesson began with a demonstration of the sperm and oocyte models. Next, the models representing the stages of fertilisation and cleavage were presented. At this point, the students were asked a few questions, such as: "Where does the cleavage process take place?". The answer received was "in the fallopian tube". Next, the model of the Female Reproductive System was presented and left on the teacher's desk for the students to identify the structures that make it up.

According to Krasilchik (2004), students retain little information during lectures and the use of educational resources favours the learning of new knowledge obtained during lectures. So whenever a new model was shown, the students were asked what it represented and what structures could be observed in each of the models, always associating them with the subjects covered during the lecture. Most of the time, the excited students answered correctly. Often, when the answers given were different, there were disputes between the students as to which answer was correct. The students were happy when they got the names of the structures right and when they had any doubts, they tried hard to remember the pictures on the slides. On several occasions, the students compared the models with the figures on the slides.

After looking at the models, many students asked how the models were produced and suggested that there should be lessons in which they could build the materials themselves. Although this activity was not carried out during the course of this work, it is undoubtedly a useful resource for teachers, who can offer greater richness to their lessons in a very simple way.

Teachers should realise that educational resources are not just those contained in a laboratory, but also, more broadly, the most varied components of the environment that can provide stimulation for both students and teachers in the various moments that involve teaching and learning. These components can serve as alternative didactic or paradidactic resources in the classroom, representing an excellent way of enriching biology lessons, even making up for the lack of laboratories in schools, since this is not the only source of teaching resources. It's obvious that experimentation should be part of school life, but it shouldn't be restricted to laboratory activities, in which students are given a recipe to follow down to the smallest detail and whose results are already known in advance. Often, simple experiments that can be carried out at home, in the schoolyard or in the classroom with everyday materials mobilise students and contribute to the development of a wide range of skills (Brasil, 1998). The teacher should therefore be aware that there is a huge diversity of resources available in everyday life, which can be used by the teacher to propose different practical lessons and improvise new teaching materials.

4.3. Game application

In general, games can be seen in society as an instrument that favours physical, cognitive, affective, social and moral development, which has two functions: a) the ludic, which provides fun and pleasure, and b) the educational, which teaches, becoming a complement to knowledge (Kishimoto, 1996). Seen in terms of its educational function, the game is seen as a material that facilitates knowledge, where content is learnt in a pleasurable and playful way.

A game can be defined as any interaction between players within a defined set of rules (Vila & Santander, 2003), which must be accepted, transformed or even destroyed when questioned by the players (Brougère, 1998). The dynamic nature of games is the main difference between them and other educational materials in the teaching-learning process, because as well as aiding learning, they provide meaningful interaction between participants. According to Rocco (1999), group dynamics and games are capable of stimulating the expression of ideas and group exchange. Games, as a methodological proposal, represent playfulness as the foundation of culture and its forms of transmission, as well as rescuing dialogue between young people and educators. The use of playful and interactive materials encourages a cognitive questioning process, favouring reflection and the acquisition of knowledge.

According to the PCNEMs, games and play are very valuable elements in the process of appropriating knowledge. They enable the development of skills in the areas of communication, interpersonal relationships, leadership and teamwork, using the relationship between co-operation and competition in a formative context. Games provide the stimulus and environment that favours the spontaneous and creative development of students and allows teachers to broaden their knowledge of active teaching techniques, develop personal and professional skills and stimulate students' ability to communicate and express themselves. However, the use of games as a pedagogical tool is not restricted to working only with ready-made games in which the rules and procedures are already determined, but mainly to creating games related to the themes discussed in the classroom context. With this in mind, this game was developed in line with the subjects presented during the lecture on the Female Sexual

Cycle.

The game took place after the presentation of the models. The students were divided into groups according to their affinity. Once again, the students were very agitated at the start of the game, which meant that it took a long time to organise the class into groups. The cards were distributed to the groups and the rules of the game were explained. In some groups, there were small discussions to decide who would get the answer card and the mini-cards. As soon as the mini cards were handed out, all the groups realised that they formed a puzzle. The game didn't start until all the rules were clear. There was no difficulty in understanding them, since all the students knew the rules of the Bingo game.

During the game, all the students remained very attentive when a question was asked. Most of the questions were answered without too much difficulty and without any help, so it can be concluded that the students were able to understand the lecture. In some of the more complex questions where it took a while to get an answer from the students, a few hints were enough for them to answer correctly. In this case, the students often couldn't remember the correct answer and looked up the correct name on the answer sheet or referred to a particular structure, making comparisons like: "that little ball", "the layer on the outside", "that thing that looks like a tube".

At the start of the game, the students were wary of inviting or expelling any members of the group, but as soon as one group expelled a member, all the other groups began to change their members. Only the winning group remained with the same members throughout the game.

Some groups couldn't find a way to build the puzzle. The winning group (figure 9g), as well as being the first group to fill in the whole card, managed to form the puzzle correctly (figure 9h). Another group assembled the puzzle even before filling in the card, making it easier to assemble during the game. The puzzle was used in this game because, in addition to its playful nature, it stimulates visual perception, spatial reasoning, the association of ideas and awakens the student's attention and concentration (Antunes, 2005).

At the end of the game, the two groups that had managed to put the puzzle together correctly were asked to explain what the figure represented in the Female Sexual Cycle. The two groups had no difficulty recognising and explaining to the rest of the room what the figures represented.

During the game, as well as learning, it was possible to observe that there was interaction between the students, with co-operation within the groups and competition between them. According to Vygotsky (1984), learning is fundamentally a social experience, of interaction through language and action, so that interaction must provide a community of learning, discourse and practice in such a way as to produce meanings, understanding and critical action, to exercise cooperative learning and autonomy, to ensure the centrality of the individual in the construction of knowledge and to enable cognitive, affective and action results. In this context, it was possible to observe at least three student positions within the groups: students who took the lead in the group, so that they were never expelled; students who actively collaborated in the group's victory and were rarely expelled; students who remained passive in the decisions, and were often expelled from the group.

Most of the students had no problem going to one group or another, but some of them refused to take part in a particular group and made their group choose another to receive them. One of the students couldn't stay in any of the groups for long, as he was constantly asked to join the group that filled in the rows or columns of the card. The winning group, which remained with the same members, refused to receive other members.

Of all the students, one stood out because of his participation and interest in all the activities. This student, who was always the first to answer, was expelled from his group and refused to take part in any other group, deciding to leave the game. Nevertheless, he remained by my side, answering all the questions correctly and handing out the sweets as the groups completed their cards. In agreement with Brougère (1998), this activity provided an opportunity to test and observe the students, because during a game, manifestations always occur spontaneously.

Although the game lasted more than an hour, the students remained excited until it was over, when one group had completed the whole card and still managed to put the puzzle together correctly.

Through this activity, it was possible to see that the game has numerous qualities and benefits. As well as working as a method for verifying learning, it was observed that several questions that were not clarified during the lecture could be better understood during the game. According to Kishimoto (1996), some knowledge acquisition processes are facilitated when they take the form of games. The use of interactive educational games has emerged as an alternative capable of bringing a significant gain in quality to the teaching-learning process. Games offer the opportunity for interactivity, as well as rescuing the playful and pleasurable aspect of learning. According to Botelho (2004), games can be used to train operational skills, raise awareness and reinforce motivation, develop perception, train communication and cooperation, integrate and apply learnt concepts in practice and even assess learning.

In short, games can become efficient tools, as they entertain and motivate, thus facilitating learning, as they increase the capacity to retain what has been taught (Campos et al., 2003). In addition, games develop the brain's cognitive structures, facilitating the development of new skills such as observing and identifying, comparing and classifying, conceptualising and relating, as well as developing creativity, perseverance, sociability and self-confidence (Silveira & Barone, 1998).

A well-structured and correctly applied game provides very rich results in terms of assimilating or reformulating concepts (Militão, 2000). On this basis, the teacher, together with the students, can create the most diverse activities, based on existing games, with surprising results. According to Kishimoto (1996), because games are free from pressure and evaluation, they create a climate of freedom that is conducive to learning, stimulating interest, discovery and the acquisition of knowledge.

All of this could be seen during the implementation of this activity, as the game developed here awakened in the students co-operation, respect for the rules, a sense of responsibility and leadership in the group, personal initiative, teamwork and decision-making, as well as an interest in learning.

Figure 9: Lecture, application of models and game. A) Slide being projected on the blackboard. Note the various defects in the blackboard that hindered the projection of the illustrations. **B-F)** Students from the 1st year of secondary school during the exhibition of the models. **G)** The winning group in the game. **H)** Puzzle put together by the winning group.

4.4. Evaluation

Assessment is an important tool for teachers because it allows them to check student performance, which is a reflection of the effectiveness of teaching. It can therefore be said that student performance mirrors the work carried out by the teacher, so by assessing their students, the teacher is also assessing their own work (Haidt, 2001). That said, to finalise the lesson, the students were given a test containing 10 multiple-choice questions and one open-ended question. As time was running out, the students were asked to answer only the open question, writing down their opinions on the lesson and the teaching resources used. However, they insisted and wanted to answer all the questions without worrying about time. So, during the test, they didn't bother copying the answers of other classmates either and did all the questions individually. This assessment is analysed in Figures 10 and 11.

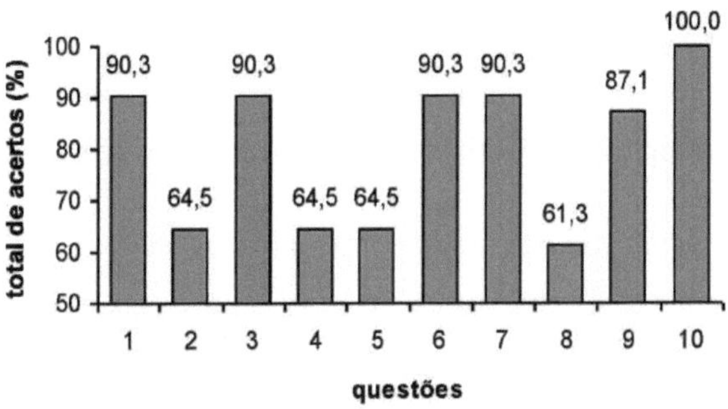

Figure 10: Percentage of correct answers to each question

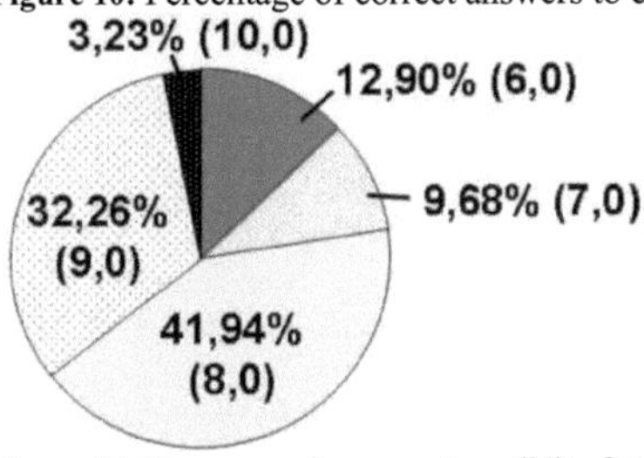

Figure 11: Frequency in percentage (%) of students' grades (in brackets)

By analysing the results of the assessment, it was possible to conclude that the students were able to assimilate the new knowledge, so that learning took place in a significant way, since the class performance was very good (figure 11), since more than 50% obtained a mark equal to or higher than 8.0 (3.23% of marks were 10.0 + 32.26% of marks were 9.0 + 41.94% of marks were 8.0, totalling 77.43% of marks equal to or higher than 8.0). Some questions were more difficult, but no question had an error rate of more than 40 per cent (figure 10). Others that were considered easier had a success rate of up to 100 per cent (figure 10).

Considering the answers given by the students in the open question, it was possible to conclude that they all liked the methods used during the lesson and that there was effective learning. Some excerpts written by the students are transcribed below:

"This class was different and enjoyable. I learnt a lot. If I could have this kind of class every week, I'd make a point of staying to learn more. The game was very creative, we played and *learnt at the same time."*

"This teaching method is more effective than the common one used: the teacher writes on the blackboard and the student copies. If classes were like this, practical, students would learn more. It's simple but effective. I'd like to learn *more this way.*

"With these materials I could understand the explanation more clearly, I could understand everything, I really liked it. I think it's very good to have classes like this, nowadays it's very important and using films, models, slides, it's *much better than the teacher just talking. "*

"Even if I didn't understand any of the subject, the game made it very easy and it was a fun way to learn. As well as having a laugh, you *learn."*

"With the model you understand how various things work and the game is very enjoyable and interactive. It was a way of putting into practice what we learnt during the lesson

According to Vasconcellos (1995), teachers cannot wait for the school to provide an ideal classroom with few students and an equipped laboratory to arouse their students' interest. It's up to the teacher to deal with these and other problems presented by the educational system, so that it's possible to arouse the student's interest during the teaching-learning process. To do this, it's enough to include didactic activities in the syllabus that have playful and enjoyable aspects, which can be created by the teacher themselves.

Once the process had been finalised, a letter was sent to the school's headmistress, thanking her for the opportunity and the facilities provided for this work (Appendix IV).

CHAPTER 5

Considerations

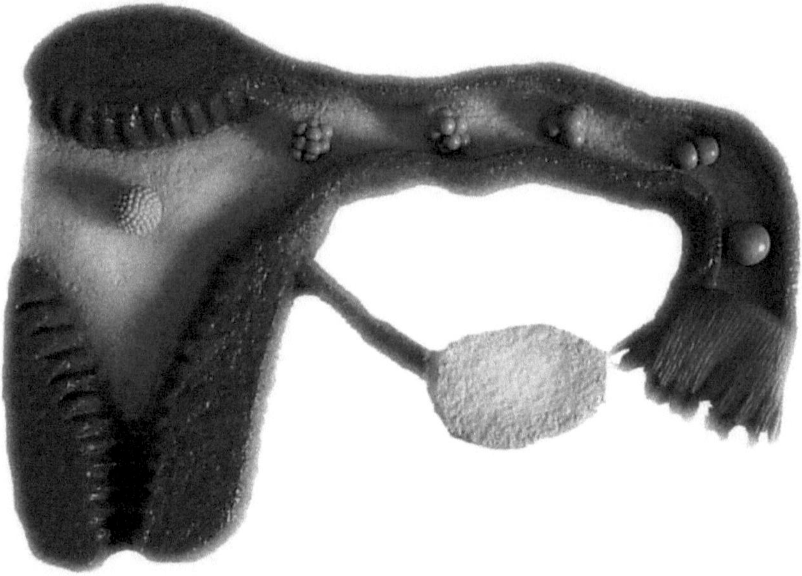

Teachers won't always find the ideal school at their disposal that offers them adequate resources to carry out the desired activities during their lessons. Teachers need to realise that their lessons should not be restricted to lectures or to the resources offered by the school, which are often unavailable, absent or in poor condition. A good example of this is the school where this work was carried out, which didn't have an overhead projector and its use was only made possible by the loan from the Morphology Department of the Biosciences Institute, UNESP, Botucatu-SP Campus. This was made easier by the fact that it was linked to a university. However, teachers in primary and secondary schools, especially in the public sector, don't always have access to these materials, especially in small towns and cities in the countryside. Therefore, a lecture using a slide show would not be possible under these conditions. This further favours the use of didactic resources made from very simple and easily available materials, such as models and games that can be created by the teachers and students themselves, like the ones presented in this paper. On the other hand, this also shows the importance of socialising knowledge between the university and the school.

According to Antunes (2005) games, in their integral sense, are the most efficient means of stimulating the intelligences. The game space allows the student to do whatever they want. When engaged in a game, the individual is who they want to be, they order what they want to order, they decide without restrictions. Thanks to play, the symbolic fulfilment of the desire to be great and the longing to be free can be achieved. Socially, play imposes control over impulses and leads to the acceptance of rules. Play sets its own limits of time and space, creates order and balances rhythm with harmony. Play helps to build new discoveries, develops and enriches the personality and symbolises a pedagogical tool that makes the teacher the conductor, stimulator and evaluator of learning. Therefore, it is more than necessary for teachers to find other ways to arouse interest and learning in their students, because there is no teaching without learning and learning only happens through transformation, through the facilitating action of the teacher and through the process of seeking knowledge. Each student becomes a challenge to the teacher's competence. The student's interest becomes the driving force behind the learning process and the teacher must be a generator of stimulating and effective situations, discoveries and experiences that keep their students' interest constant. It is in this context that teaching materials have gained a place as the ideal tool in the learning process, insofar as they stimulate student interest.

Still from this perspective, it was possible to observe during the course of the activities that learning and student participation become much more satisfactory when resources are used that attract the student's attention and reinforce the retention of the concepts learnt. The making of models by the teacher and the students themselves are resources that can always be available in a classroom, without needing equipped laboratories or suitable rooms to carry out certain activities. These activities, as well as creating a relationship of co-operation and competition between the students, show them a new playful, enjoyable and participatory way of relating to school content, leading to a greater appropriation of the knowledge involved. Therefore, in a simple way and with a little creativity, teachers can create valuable elements in the process of appropriating knowledge, making their lessons more dynamic and attractive, involving their students in a more meaningful way in the teaching-learning process.

5.1. Proposal

Based on this experience, here is a proposal for the use of teaching resources in general and the material produced in this work.

- **When to use educational resources**

Paradidactic resources can be used when there is a possibility within the programme and when they are an efficient aid to achieving an objective within that programme. They should appear in the pedagogical proposal and always be applied by the teacher with a critical spirit in order to keep them, change them or replace them with others when it is realised that they have fallen short of the objectives (Antunes, 2005).

Thus, these resources will be valid and meaningful if used at the right time, which is determined by the student's interest, the proposed objective and the challenging nature of the teaching resources.

- **Proposal for using the educational materials produced**

The educational materials developed in this work can be used by teachers in the following situations:

Game application:

N In Human or Animal Physiology classes, when the Female Sexual Cycle is covered, as well as the following topics: female hormones, uterine cycle, ovarian cycle, gametogenesis, folliculogenesis, reproductive period, female reproductive system, feedback mechanism, constitution of the female gamete.

Application of the models:

N In Embryology lessons, when fertilisation and cleavage are covered.

N In Zoology lessons, when Reproduction is covered.

N In Genetics lessons, when Gametogenesis is covered.

The educational material produced here was publicised in high schools in the city of Botucatu-SP through personal contact and explanatory folders:

- E.E. "Prof. Américo Virginio dos Santos"
- E.E. "Dr Armando de Salles Oliveira"
- E.E. "Cardoso de Almeida"
- Dom Lúcio Antunes de Souza" school
- Prof Pedro Torres" school
- Prof José Pedretti Neto" school
- Our Lady of Lourdes Archdiocesan College - La Salle
- Santa Marcelina College
- Colégio Objetivo
- Arrow College
- Anglo College

The material is currently being disseminated to schools in the city of Alfenas, MG.

This material is available for loan at the following universities:

- Department of Morphology, Botucatu Biosciences Institute, UNESP, Botucatu-SP Campus. Professor responsible: Dr Maria Dalva Cesario

- Department of Cell and Developmental Biology, Institute of Biomedical Sciences, UNIFAL, Campus Sede, Alfenas-MG. Professor responsible: Dr Talita Sarah Mazzoni

Bibliographical references

ANTUNES, C. *Games for stimulating multiple intelligences.* 13ª edition, Petrópolis: Vozes,

2005. 295p.

BOTELHO, L. *Educational games applied to e-learning*. 2004. Available at: <http://www.elearningbrasil.com.br/news/artigos/artigo_48.asp.> Accessed on: June 2005.

BRAZIL. MEC. *National curriculum parameters for secondary education*. Brasília, MEC/SEF, 1998. Available at: <http://www.mec.gov.br/semtec/ensmed/pcn.shtm.> Accessed on: July 2005

BROUGÈRE, G. *Play and education*. Porto Alegre: Artes Médicas, 1998. 218p.

CAMPOS, L.M.L., BORTOLOTO, T.M.; FELÍCIO, A.K.C. *The production of didactic games for teaching Science and Biology: a proposal to favour learning*. Núcleos de Ensino da Unesp, São Paulo, 2003. Available at: <http://www.unesp.br/prograd/PDFNE2002/aproducaodejogos.pdf> Accessed on: March 2017.

CAMPOS, M.C.C., NIGRO, R.G. *Didática de ciências: o ensino-aprendizagem como investigação*. São Paulo: FTD, 1999. 190p.

COLL, C., MARTÍN, E., MAURI, T., MIRAS, M. ONRUBIA, J., SOLÉ, I., ZABALA, A. *O construtivismo na sala de aula*. 6th edition, São Paulo: Ática, 2003. 221p.

DELIZOICOV, D., ANGOTTI, J.A., PERNANBUCO, M.M. *Ensino de ciências: fundamentos e métodos*. São Paulo: Cortez, 2002. 364p.

GARCIA, S.M.L., FERNÁNDEZ, C.G. *Embryology*. 3rd edition, Porto Alegre: Artmed Editora Ltda, 2011. 668p.

HAIDT, R.C.C., *Curso de didática geral*. 7th edition, São Paulo: Editora Ática, 2001. 327p.

HARLAN, J.D., RIVKIN, M.S. *Ciências em educação infantil: uma abordagem integrada*. 7th edition, Porto Alegre: Artmed, 2002. p.21-64

KISHIMOTO, T.M., *Game, toy, play and education*. São Paulo: Cortez, 1996. 183p.

KRASILCHIC, M. *Prática de ensino de biologia*. 4ª edition, São Paulo: Editora da Universidade de São Paulo, 2004. p. 85-87

KRASILCHIC, M. *O professor e o currículo das ciências*. São Paulo: Editora da Universidade de São Paulo,1987. 80p.

KRASILCHIC, M. *Prática de ensino de biologia.* São Paulo: Harper & Row do Brasil Ltda, 1983. 203p.

MILITÃO, A. *Jogos, dinâmicas & e vivências grupais.* Rio de Janeiro: Qualitymark Editora, 2000. 226p.

OSSE, A.C. *Where did embryology end up?* São Paulo: Fovest - Folha de São Paulo, 31/10/2000.

ROCCO, J.J. *Sexualidade e mudanças de comportamentos: uma estratégia lúdica de prevenção da AIDS.* Rio de Janeiro: Zahar, 1999. p. 175-199

SILVEIRA, R.S., BARONE, D.A.C. *Computerised educational games using the genetic algorithms approach.* Federal University of Rio Grande do Sul. Institute of Informatics, IV RIBIE Congress, Brasilia, 1998. 13p.

VASCONCELLOS, C.S. *O planejamento como instrumento da práxis pedagógica.* São Paulo: Cadernos Pedagógicos de Libertad 1, 1995. p.30-34

VEIGA, I.C.A. *In school laboratories and workshops: the didactic demonstration. Teaching techniques: Why not?* São Paulo: Papirus, 2003.

VILA, M., SANTANDER, M. *Jogos cooperativos no processo de aprendizagem acelerada.* Rio de Janeiro: Qualitymark, 2003.

VYGOTSKY, L.S. *The social formation of the mind.* São Paulo: Martins Fontes, 1984.

ILLUSTRATIONS:

<http://www.reprodusite.hpg.ig.com.br/inde.htm> Accessed: April and May 2005

<http://www.bioscience.org/atlases/fert/embryoge.htm> Accessed: April and May 2005

<http://www.biologia.net> Accessed: April and May 2005

<www.sexualidades.com.br> Accessed: April and May 2005

<www.planetabebe.hpg.ig.com.br> Accessed: April and May 2005

<www.bebesenlaweb.com.ar/elprincipiodelavida.html.> Accessed on: April and May 2005

<http://www.biologo.com.br> Accessed: April and May 2005

EMBRYOLOGY CONTENT IN PREPARATION FOR THE LECTURE:

DEMAREST, R.J. Illustrated guide to human reproduction & fertility control. New York, USA: The Parthenon Publishing Group, 1996. 96p.

GUYTON, A.C., HALL, J.E. *Treatise on medical physiology.* 12ª edition, Rio de Janeiro: Guanabara Koogan, 1997. 1176p.

JUNQUEIRA, L.C., CARNEIRO, J. *Basic histology.* 9th edition, January: Guanabara Koogan, 1999. p. 355-388.

MOORE, K.L., PERSAUD, T.V.N. *Basic embryology.* 5th edition, Rio de Janeiro: Guanabara Koogan, 2000. 453p.

SOBOTTA, J. *Atlas of human anatomy.* 21st edition, Rio de Janeiro: Editora Guanabara Koogan S A, 2000. p. 195-204.

CHAPTER 6

Annexes

Annex I: Biscuit dough

Ingredients:

1 1 glass of white glue

2 *2* cups cornstarch

1 1 tablespoon liquid Vaseline

Preparation:

Put the glue and liquid petroleum jelly in a bowl and mix with a spoon. Gradually add the cornstarch until you can no longer stir it with a spoon. Add the rest of the starch and keep stirring the dough with your hands until the dough is soft and comes away from your hands. The dough can be used immediately. If you wish, store it in a plastic bag.

Biscuit dough should be coloured with oil-based paints, but in this work, all the dough used was coloured with gouache or painted after drying.

Annex II: Papier-mâché dough

Ingredients:

5 5 toilet rolls

water

2 250 ml vinegar

J ^ kg of gypsum

5 500 ml white glue

Preparation:

Put the water and vinegar in a bucket. Then add the toilet rolls and leave to soak for 24 hours. After this time, remove the rolls from the water and break them into small pieces until they are completely crumbled. Dry them well with a cloth. When they are very dry, put the crumbled paper in a bowl with 1 litre of water and the glue. Stir until you get a soft dough, then gradually add the plaster. Keep stirring until you have a homogeneous, mouldable mass. The dough should be used immediately after preparation.

Toilet paper can be replaced with any other type of paper, but paper that is more impermeable should be soaked for longer than 24 hours until it is completely crumbly.

Annex III: Evaluation model applied after the end of the lesson

Name: ____

1. What are the female gamete and the female gonad called, respectively?
   a) sperm and ovary
   b) ovary and ovum
   c) ovary and sperm
   d) egg and sperm
   e) egg and ovary

2. Which hormones are produced by the pituitary gland?
   a) FSH, LH and LTH
   b) progesterone
   c) chorionic gonadotrophin
   d) GnRH
   e) oestrogen

3. What hormones does the ovary produce?
   a) FSH and LH
   b) LH and LTH
   c) oestrogen and progesterone
   d) gonadotropic hormone and progesterone
   e) oestrogen and chorionic gonadotrophin

4. What are the phases of the uterine cycle?
   a) follicular phase and proliferative phase
   b) proliferative phase, secretory phase and regression phase
   c) lutein phase and regression phase
   d) follicular phase and lutein phase
   e) secretory phase and lutein phase

5. What does negative feedback mean?
   a) release of hormones
   b) temporary stoppage of hormone release
   c) hormone production
   d) stimulation of the pituitary gland
   e) none of the above

6. Which ovarian structure produces the hormone progesterone?
   a) albicans body
   b) corpus luteum
   c) Graaf follicle
   d) granular
   e) zona pellucida

7. Which female hormone is produced during pregnancy by the placenta?
   a) oestrogen

b) progesterone
c) chorionic gonadotrophin
d) FSH
e) LH

8. What is cleavage and where does it occur?
   a) mitotic divisions of the zygote and occurs in the fallopian tube
   b) mitotic divisions of the zygote and occurs in the uterus
   c) mitotic divisions of the zygote and occurs in the ovary
   d) meiotic divisions of the zygote and occurs in the uterus
   e) meiotic divisions of the zygote and occurs in the fallopian tube

9. What is the stage of the zygote that will implant in the uterine wall?
   a) morula
   b) blastocyst
   c) egg or zygote
   d) 4-cell stage
   e) 8-cell stage

10. In which organ does embryonic development take place?
   a) uterus
   b) fallopian tube
   c) ovary
   d) ovarian follicle
   e) none of the above

What do you think of the material used during the lesson (slide show, models and the game)?

____
____
____
____
____

Annex IV: Thank you letter

Botucatu, 5th October 2005.

Mrs Ilmas
Director and Vice Director
Professor Américo Virgínio dos Santos State School

Botucatu - SP

Dear Ladies,

I would like to thank you for the space you gave me to apply the material for the Biological Sciences Degree Programme at the Institute of Biosciences in Botucatu, UNESP, to students in the first year of secondary school. I am grateful for all the attention, kindness and respect with which I was received at this school, both by the management and by the teachers who gave up their classes during the period I used to apply the material. I would also like to say a special thank you to the 1st year A students, who surprised me with their interest, participation and collaboration during the activities.

Activities carried out:

- Lecture on the Female Sexual Cycle, Fertilisation and Cleavage, with slide presentation.
- Presentation of models of the Female Reproductive System, cleavage stages, male gamete, female gamete and fertilisation stages.
- Making a game about the Female Sexual Cycle and then applying a test-type assessment to check learning.

The final result was so satisfactory that it would be interesting for you and all the teachers who collaborated in this process to be informed of it, and I enclose the results of the evaluation.

Yours sincerely,

Prof Talita Sarah Mazzoni **Prof Maria Dalva Cesario**

END

Fertilisation marks the beginning of one of the most admirable phenomena in biology - the process of embryogenesis, in which the zygote develops into a new individual?

MIX
Papier aus verantwortungsvollen Quellen
Paper from responsible sources
FSC® C105338

Printed by Books on Demand GmbH, Norderstedt / Germany